The
Formation

A Cosmic journey from nothing to infinity

Manjeet Barman

Preface

Welcome, fellow seekers of knowledge, to " The Formation: A Cosmic Journey from nothing to infinity " Within the pages of this book, we embark on an extraordinary expedition, exploring the profound mysteries that have shaped our existence.

Throughout history, humanity has been gripped by a relentless curiosity about the origins of the universe and the intricate web of life on Earth. We have marveled at the stars, contemplated the vastness of space, and wondered how it all began. This book is born from that insatiable thirst for understanding and the burning desire to share the wonders of our cosmic heritage.

This book is a testament to the tireless dedication of countless scientists, thinkers, and explorers who have delved into the depths of knowledge to unravel the mysteries that surround us. Their contributions, coupled with the latest scientific discoveries, form the bedrock upon which this narrative is built.

But this book is not just about facts and figures; it is an invitation to embark on a personal journey of intellectual exploration. It invites you, dear reader, to join in the wonder, to question the mysteries, and to marvel at the beauty of our cosmic story. Together, we will challenge our perceptions, expand our understanding, and celebrate the intricate interconnectedness of the universe.

As you turn each page, I encourage you to embrace the awe that comes from contemplating our place in this vast cosmic symphony. Allow yourself to be inspired, to revel in the elegance of nature's design, and to kindle a sense of wonder that ignites your own intellectual pursuits.

So, let us embark on this voyage of discovery together, embracing the enigmatic and the unknown. May this book serve as a guiding light, illuminating the remarkable tale of the universe's birth, the formation of our Earth, and the intricate web of life that binds us all.

Welcome to " The Formation: A Cosmic Journey from nothing to infinity"

Introduction

Welcome to the captivating world of cosmic origins and the remarkable story of our own existence. In this book, " The Formation: A Cosmic Journey from nothing to infinity," we embark on an awe-inspiring exploration of the grand tapestry that encompasses the origins of the universe, the birth of galaxies, the formation of our planet, and the emergence of life as we know it.

As humans, we have an innate curiosity about the origins of everything around us. From the vast expanse of the cosmos to the intricate ecosystems that thrive on Earth, we yearn to comprehend how it all came to be. This book is a comprehensive guide that brings together the latest scientific discoveries, compelling theories, and profound insights into a cohesive narrative that illuminates our cosmic heritage.

In the chapters that follow, we will embark on a journey through time and space, unraveling the mysteries of the universe's birth and the forces that shaped it. We will delve into the cataclysmic events that forged galaxies, stars, and planets, including our own. By tracing the cosmic timeline, we will witness the intricate dance of particles and the extraordinary processes that sculpted the universe into its present form.

This book is not merely a compilation of scientific facts and theories; it is an invitation to embark on a captivating intellectual voyage. We will encounter the groundbreaking work of visionary scientists, the complexities of astrophysics, geology, and biology, and the profound philosophical questions that arise when contemplating our place in the universe.

Whether you are an avid science enthusiast, a student seeking knowledge, or a curious mind longing for a deeper understanding of the cosmos and our place within it, "From Cosmos to Earth" aims to satisfy your thirst for knowledge. By synthesizing cutting-edge research with accessible language, we strive to bridge the gap between scholarly exploration and general readership, making complex concepts comprehensible without compromising accuracy.

So, let us venture forth together, ready to explore the magnificent origins of the cosmos, the intricate tapestry of our planet, and the astonishing emergence of life. The journey awaits, and the mysteries of the universe beckon us to uncover the profound secrets that lie within.

Contents

About the Universe

The universe is the entire expanse of space and time, including all matter and energy, and the physical laws and constants that govern them. Approximately 13.8 billion years ago, a massive explosion called the Big Bang is believed to have caused it. Cosmologists and scientists continue to study and debate the structure and composition of the universe, which is still expanding and evolving. Galaxies, stars, planets, and other celestial bodies make up the universe, as well as vast amounts of dark matter and dark energy that are still unknown. The study of the universe is called cosmology, which encompasses a wide range of disciplines including physics, astronomy, and astrobiology.

The universe is made up of a variety of elements, including:

➤ Matter: Matter is the physical substance that makes up all objects in the universe, including stars, planets, and galaxies. It is made up of atoms, which are the building blocks of all elements.

- ➤ Energy: Energy is the ability to do work and is present in various forms, including electromagnetic radiation (such as light and radio waves), thermal energy, and kinetic energy (the energy of motion).
- ➤ Dark matter: Dark matter is a hypothetical form of matter that is thought to make up approximately 85% of the universe's total matter. It does not emit, absorb or reflect any electromagnetic radiation, making it invisible to telescopes.
- ➤ Dark energy: Dark energy is a hypothetical form of energy that is thought to be responsible for the acceleration of the expansion of the universe.
- ➤ Cosmic microwave background radiation: The cosmic microwave background radiation is a faint glow of light that fills the universe and is thought to be the afterglow of the Big Bang.
- ➤ Gravitational waves: Gravitational waves are ripples in the fabric of space-time caused by massive acceleration of objects, such as the merging of two black holes.

The universe is a vast and complex place, and scientists are still working to understand all of its intricacies. However, some key principles that are known include the laws of physics, such as gravity and electromagnetism, which govern the behavior of matter and energy. The universe is also thought to be expanding, and it is believed that all of the matter and energy in the universe was once concentrated in a single point known as the Big Bang. This explosion is thought to have occurred around 13.8 billion years ago, and the universe has been expanding ever since. Additionally, scientists believe that the universe is made up of dark matter and dark energy, which cannot be directly observed but can be inferred through their effects on the behavior of visible matter.

One of the most important principles in understanding the universe is the theory of general relativity, proposed by Albert Einstein in 1915. This theory describes the behavior of gravity as the curvature of space-time caused by the presence of matter or energy. This theory has been extensively tested and has been found to be extremely accurate in explaining the behavior of large objects, such as planets and stars.

Another key principle in understanding the universe is the theory of quantum mechanics, which describes the behavior of particles at the subatomic level. This theory has been found to be extremely accurate in explaining the behavior of small objects, such as atoms and subatomic particles. However, the theories of general relativity and quantum mechanics are not fully compatible with each other, and scientists are still working to develop a theory that can explain the behavior of both large and small objects.

The Big Bang theory is the most widely accepted explanation for the origins of the universe. According to this theory, the universe began as an extremely hot and dense point known as a singularity. This singularity expanded and cooled, eventually leading to the formation of atoms and the formation of stars and galaxies. The Big Bang theory also predicts that the universe is still expanding today, and that the rate of expansion is increasing.

One of the key pieces of evidence for the Big Bang theory is the cosmic microwave background radiation, which is a faint glow of light that can be observed in all directions. This radiation is thought to be the afterglow of the Big Bang, and its temperature is extremely uniform, which is consistent with the predictions of the Big Bang theory.

Another important aspect of the universe is the presence of dark matter and dark energy. These entities cannot be directly observed, but their presence can be inferred through their effects on the behavior of visible matter. Dark matter is thought to make up about 85% of the matter in the universe, and it is believed to be responsible for the formation of large-scale structure, such as galaxy clusters and super clusters. Dark energy, on the other hand, is thought to make up about 70% of the energy in the universe and is believed to be responsible for the acceleration of the universe's expansion.

The study of the universe also involves the investigation of the properties and behavior of celestial objects such as stars, galaxies, and planets. Stars are formed when clouds of gas and dust collapse under the influence of gravity. Once formed, stars generate energy through nuclear fusion reactions, which take place in their cores. As stars age, they evolve through different stages and eventually die, resulting in the formation of new generations of stars.

Galaxies, on the other hand, are large collections of stars, gas, and dust held together by gravity. There are several different types of galaxies, including spiral, elliptical, and irregular. Scientists are still trying to understand how galaxies form and evolve over time.

Future of the Universe

The future of the universe is a topic of much debate among scientists and cosmologists. The current leading theory is that the universe will continue to expand at an accelerated rate due to the influence of dark energy. This expansion is predicted to continue indefinitely, with the average temperature of the universe eventually reaching a state known as "heat death," where all matter is evenly distributed and no energy can be transferred.

However, there are other possibilities that have been proposed as well. One of these is the "big crunch" theory, in which the expansion of the universe is eventually reversed and all matter is pulled back into the singularity from which it originated. Another possibility is the formation of a "big rip," in which the expansion of the universe accelerates so rapidly that it tears apart all matter, including atoms and subatomic particles.

It's also important to note that it's not entirely clear what will happen to the universe in the far future, as the theories that scientists have come up with so far have many uncertainties and gaps.

There are also several ongoing efforts and projects that aim to shed more light on the fate of the universe, such as the Euclid space telescope, which will study the properties of dark matter and dark energy to help constrain the future of the universe, and the James Webb Space Telescope, which will study the early universe and help to understand how the first galaxies and stars were formed.

It's worth noting that the universe is vast and ancient, and our understanding of it is still limited. Many new discoveries and advances in technology are likely to be made in the future, which will help scientists to better understand the universe and its future.

Origin of the Universe.

It is a complex and highly debated topic in physics and cosmology to determine the origin of the universe. There are several leading theories that attempt to explain how and why the universe came into existence.

One of the most widely accepted theories is the Big Bang theory. According to this theory, the universe began as an incredibly hot and dense singularity around 13.8 billion years ago. This singularity rapidly expanded and cooled, eventually leading to the formation of subatomic particles, atoms, and eventually stars and galaxies. The Big Bang theory is supported by a wide range of observational evidence, including the cosmic microwave background radiation, the large scale

structure of the universe, and the abundance of light elements.

Another theory that has gained significant attention in recent years is the theory of cosmic inflation. This theory proposes that the universe underwent a period of extremely rapid expansion in the first fraction of a second after the Big Bang. This expansion would have smoothed out any initial inhomogeneities and could explain certain features of the cosmic microwave background radiation.

The multiverse theory also proposes that our universe is just one of many universes, or multiverse, that exist and the big bang is just the beginning of one of these universes.

The string theory and the loop quantum gravity are also theories trying to explain the origin of the universe; however, they are not yet supported by experimental evidence.

In summary, while the exact origins of the universe are still not fully understood, the Big Bang theory and the theory of cosmic inflation are currently the most widely accepted explanations. These theories are supported by a wealth of observational evidence, but they are still subject to ongoing research and refinement.

Formation of the Universe.

The formation of the universe is a topic that is still being studied and understood by scientists and cosmologists. The current leading theory for the formation of the universe is the Big Bang theory. According to this theory, the universe began as a singularity, an incredibly hot and dense point, around 13.8 billion years ago. This singularity rapidly expanded and cooled, eventually leading to the formation of subatomic particles, atoms, and eventually stars and galaxies.

As the universe expanded and cooled, subatomic particles began to form, including protons and neutrons. These particles eventually combined to form atoms, primarily hydrogen and helium. These atoms then came together to form stars and galaxies through the process of gravitational attraction.

As the universe continued to expand, small initial density fluctuations in the distribution of matter led to the formation of larger structures such as galaxy clusters and super clusters. Through the process of gravitational instability, matter concentrated in certain regions, forming the large scale structure of the universe we observe today.

The formation of the first stars and galaxies marks the end of the cosmic "dark ages" and the beginning of the "cosmic dawn". These first luminous objects reionized the neutral hydrogen present in the universe and cleared the way for the formation of subsequent generations of stars and galaxies.

It's important to note that the formation of the universe is an ongoing process and new discoveries and observations continue to refine our understanding of the early universe. Theories such as cosmic inflation and the multiverse also propose alternative explanations for the formation of the universe.

In summary, the current leading theory for the formation of the universe is the Big Bang theory, which states that the universe began as a singularity and expanded and cooled to form subatomic particles, atoms, stars, and galaxies. The formation of the universe is thought to be an ongoing process, with new discoveries and observations continuing to refine our understanding of the early universe.

The Big-Bang Theory & its Evidence.

The Big Bang theory is a scientific explanation for how the universe began. It states that about 13.8 billion years ago, there was nothing - no stars, no galaxies, no atoms, nothing. But then, in a split second, an incredibly hot and dense point called a singularity began to expand rapidly. This expansion is thought to have occurred in a fraction of a second and it caused the universe to cool down and form subatomic particles, atoms, and eventually stars and galaxies.

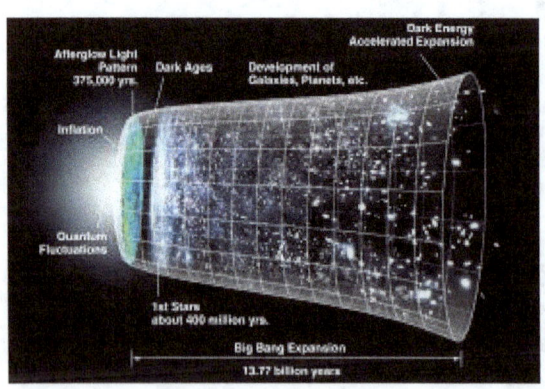

One of the key pieces of evidence that supports the Big Bang theory is the cosmic microwave background radiation. This is a faint glow of light that fills the universe and has a temperature of about 2.725 degrees above absolute zero. This radiation is thought to be the afterglow of the intense heat that was present during the early stages of the Big Bang. Scientists have been able to detect this radiation using special telescopes and have found that it is extremely uniform, which supports the idea that it is a remnant of the intense heat present in the early universe.

Another piece of evidence that supports the Big Bang theory is the large scale structure of the universe. Scientists have been able to map the distribution of galaxies in the universe and have found that they are arranged in a "cosmic web" of filaments and voids. This structure is thought to have formed as a result of the initial density fluctuations present in the early universe, which were amplified by the expansion of the universe.

The abundance of light elements is also an evidence of the Big Bang theory. During the early stages of the universe, the universe was too hot for atoms to form. However, as it expanded and cooled down, protons and neutrons began to form and eventually combined to form light elements like hydrogen and helium. The relative abundance of these elements in the universe is consistent with what we would expect if the Big Bang had occurred.

The Big Bang theory also predicts that the universe should be expanding. This prediction was first made by Edwin Hubble in the 1920s and has been confirmed by numerous observations. Scientists have been able to measure the distances to faraway galaxies and have found that they are moving away from us at an accelerating rate. This is thought to be caused by the expansion of the universe and is consistent with the predictions of the Big Bang theory.

Another prediction of the Big Bang theory is the existence of dark matter and dark energy. The Big Bang theory predicts that the universe should contain a large amount of invisible matter and energy that cannot be detected by telescopes. Scientists have inferred the existence of dark matter and dark energy by observing their effects on the motion of galaxies and the large scale structure of the universe.

In summary, the Big Bang theory is a widely accepted explanation for the origin of the universe. It states that the universe began as a hot and dense singularity that rapidly expanded and cooled, eventually leading to the formation of stars and galaxies. The evidence for the Big Bang theory includes the cosmic microwave background radiation, the large scale structure of the universe, the abundance of light elements, the expansion of the universe, and the existence of dark matter and dark energy. While the Big Bang theory is still being refined and updated as new discoveries are made, it remains the most widely accepted explanation for the origin of the universe.

Geometry of Universe

The geometry of the universe refers to the overall shape or structure of the universe as a whole. The three main possibilities are a flat universe, an open universe, and a closed universe.

➢ In a flat universe, the overall curvature of space is zero, and parallel lines will remain parallel forever. The density of matter in a flat universe is exactly at the critical density required for the universe to be flat.

➢ In an open universe, the overall curvature of space is negative and parallel lines will eventually diverge. The density of matter in an open universe is less than the critical density.

➢ In a closed universe, the overall curvature of space is positive and parallel lines will eventually converge. The density of matter in a closed universe is greater than the critical density.

The current scientific consensus is that the universe is flat or nearly flat. This is supported by various observations such as cosmic microwave background radiation, large scale structure of galaxy distribution and weak lensing.

About Galaxy.

A galaxy is a massive collection of stars, gas, dust, and dark matter held together by gravity. It can contain anywhere from a few million to trillions of stars. The Milky Way, the galaxy we live in, is thought to contain around 100 billion stars.

Galaxies come in many shapes and sizes, including spiral, elliptical, and irregular. Spiral galaxies, like the Milky Way, have a central bulge surrounded by a disk with spiraling arms. Elliptical galaxies are round or oval in shape, and contain a smooth distribution of stars. Irregular galaxies lack a well-defined shape or structure.

The study of galaxies is called galactic astronomy, and it is a key area of modern astrophysics. Scientists study galaxies to learn more about the formation and evolution of the universe, as well as the behavior and interactions of stars, gas, and other components within a galaxy.

Galaxies are thought to have formed through the gravitational collapse of massive clouds of gas and dust in the early universe. This process led to the formation of the first stars and galaxies, which then went on to grow and merge with other galaxies over time.

One important concept in galactic astronomy is the concept of dark matter. This is a type of matter that does not emit, absorb, or reflect light, but can still be detected through its gravitational effects on visible matter. Dark matter is thought to make up the majority of the mass in the universe, and plays a critical role in the formation and evolution of galaxies.

Another key aspect of galactic astronomy is the study of supermassive black holes, which are thought to reside at the centers of most galaxies. These black holes can have masses millions to billions of times that of our sun, and are thought to play a critical role in shaping the structure and behavior of galaxies.

In summary, galaxies are massive collections of stars, gas, dust, and dark matter held together by gravity. They come in a variety of shapes and sizes, and the study of galaxies is an important area of astrophysics that helps us learn about the formation and evolution of the universe, as well as the behavior of its various components.

Types of Galaxy

Galaxies, the vast cosmic structures that contain stars, gas, dust, and other celestial objects, come in various types. Here are some of the main types of galaxies found in the universe:

➢ Spiral Galaxies: Spiral galaxies feature a prominent disk-shaped structure with spiral arms extending outward from a central bulge. These arms contain young stars, gas, and dust. Our Milky Way galaxy is a prime example of a spiral galaxy.

➢ Elliptical Galaxies: Elliptical galaxies are characterized by their smooth and elliptical shape, lacking the distinct spiral arms found in spiral galaxies. They typically contain older stars and less interstellar matter. Elliptical galaxies come in a range of sizes, from small to massive giants.

➢ Irregular Galaxies: Irregular galaxies lack a defined shape and structure, often displaying chaotic and irregular features. They can have a mix of young and old stars, as well as regions of active star formation. Irregular galaxies are commonly observed in close interactions or mergers between galaxies.

➢ Lenticular Galaxies: Lenticular galaxies, also known as S0 galaxies, possess characteristics of both spiral and elliptical galaxies. They have a disk-like structure similar to spirals but lack the well-defined spiral arms. Lenticular galaxies contain older stars and exhibit less ongoing star formation.

➢ Dwarf Galaxies: Dwarf galaxies are relatively small in size compared to other galaxy types. They can be found in various shapes, including irregular, elliptical, or dwarf spiral. Dwarf galaxies often orbit larger galaxies and are thought to be the building blocks for larger galactic structures.

➢ Barred Spiral Galaxies: Barred spiral galaxies feature a central bar-like structure that extends through the galactic center, connecting the spiral arms. These arms can appear more elongated and pronounced compared to regular spiral galaxies.

➢ Ultra-Compact Dwarf Galaxies: Ultra-compact dwarf galaxies are exceptionally dense and compact, containing a high concentration of stars in a relatively small volume. They are considered to be the most compact known galaxies.

➢ Ring Galaxies: Ring galaxies exhibit a distinctive ring-like structure surrounding a central core. These rings are typically formed by a collision or close encounter between two galaxies, resulting in a shockwave that triggers the formation of new stars in a circular pattern.

These are just a few examples of the many types and classifications of galaxies found throughout the universe. Each type offers unique insights into the cosmic processes and evolutionary paths that shape the vastness of our universe.

Formation of Galaxies.

The formation of a galaxy is a complex and ongoing process that is not fully understood by scientists. However, current theories propose that galaxies formed through the process of hierarchical structure formation, where small structures merge to form larger structures.

The first step in the formation of a galaxy is the collapse of a large cloud of gas and dust. This collapse is triggered by gravity and is triggered by density fluctuations in the early universe. The collapse causes the cloud to heat up and begin to form stars. As more and more gas and dust are pulled into the center of the cloud, the density and temperature increase, causing the formation of more stars. This process continues until the cloud becomes a dense, hot, and luminous protogalaxy.

As the protogalaxy continues to collapse, the newly formed stars begin to orbit the center of the galaxy. Over time, these orbits become more organized, forming a rotating disk of stars. The disk is also surrounded by a halo of dark matter, which is thought to make up the majority of the mass of the galaxy. The dark matter acts as a gravitational glue, holding the galaxy together and providing the necessary mass to keep the stars in orbit.

As the galaxy continues to evolve, it begins to form distinct structures such as spiral arms, bulges, and bars. These structures are thought to form as a result of gravitational interactions between the stars, gas, and dust in the galaxy. For example, the spiral arms in a spiral galaxy are thought to form as a result of density waves that travel through the disk.

As the galaxy continues to evolve, it also begins to accrete smaller galaxies through mergers and accretion. These mergers can have a significant impact on the structure of the galaxy, and can lead to the formation of elliptical galaxies. The mergers can also trigger bursts of star formation, as the gas and dust in the smaller galaxy is compressed and heated by the collision.

Over time, the galaxy will continue to evolve, with new stars being formed and old stars dying. The galaxy will also continue to interact with its environment, through the accretion of intergalactic gas and the influence of nearby galaxies.

In summary, the formation of a galaxy is a complex and ongoing process that is thought to begin with the collapse of a large cloud of gas and dust. As the cloud collapses, it heats up and begins to form stars, which eventually form a rotating disk. Over time, the galaxy will continue to evolve, forming distinct structures and accreting smaller galaxies. While the current theories provide a general explanation of how galaxies formed, more research and observations are needed to fully understand this process.

A galaxy is a massive collection of stars, gas, dust, and dark matter that are held together by gravity. The main components of a galaxy include:

➢ Stars: Galaxies are made up of billions of stars of all different sizes and ages. Some galaxies have more stars than others, and the number of stars in a galaxy can be used to classify it.

➢ Gas and Dust: Galaxies also contain large amounts of gas and dust. This material is found between the stars and is important for the formation of new stars. The gas and dust in a galaxy also helps to absorb and scatter light, making the galaxy appear more colorful.

➢ Dark Matter: Galaxies contain large amounts of dark matter, which is a type of matter that does not emit, absorb, or reflect electromagnetic radiation, making it invisible to telescopes. Dark matter makes up the majority of the mass in a galaxy and is responsible for holding the galaxy together.

➢ Black Hole: Some galaxies have supermassive black holes at their center. These extremely dense objects are formed from the collapse of massive stars and can have masses equivalent to billions of suns.

➢ Globular Clusters: Galaxies also contain globular clusters, which are groups of hundreds of thousands of stars that are tightly bound together by gravity. These clusters are usually found in the halo of a galaxy and are some of the oldest known structures in the universe.

➢ Planetary Systems: Galaxies also contain planetary systems, which are groups of planets and other small bodies orbiting around a star.

➢ Halo: Some galaxies also have a halo, which is a large, spherical region of diffuse gas and dark matter that surrounds the main body of the galaxy.

➢ Spiral Arms: Some galaxies, like the Milky Way, have spiral arms, which are regions of stars and gas that spiral out from the center of the galaxy.

➢ Bulge: Some galaxies have a bulge at the center, which is a region of stars that are more tightly packed than those in the disk.

➢ Cosmic web: Galaxies are not isolated, they are connected to each other through the cosmic web, a vast interconnected network of dark matter, gas, and galaxies.

Note that, not all galaxies have all the components mentioned above, and some galaxies may have additional components not listed here.

About Milky Way Galaxy

The Milky Way galaxy has a rich history dating back billions of years. The earliest observations of the galaxy can be traced back to ancient civilizations, such as the Egyptians and the Greeks. They observed the galaxy as a band of light

stretching across the night sky and named it after its appearance.

During the Renaissance, scientists such as Galileo and Kepler made important contributions to our understanding of the galaxy. Galileo was the first to use a telescope to observe the night sky and discovered that the

Milky Way was made up of a vast number of stars. Kepler, on the other hand, discovered that the Milky Way was just one of many galaxies in the universe.

In the 19th century, William Herschel made significant contributions to our understanding of the Milky Way by mapping the distribution of stars and determining that the galaxy had a spiral shape. His son, John Herschel, also made important contributions by using telescopes to observe the galaxy in greater detail and discovering that it contained nebulae and clusters of stars.

In the 20th century, our understanding of the Milky Way continued to evolve as new technologies, such as radio telescopes and space-based observatories, were developed. These new tools allowed scientists to study the galaxy in unprecedented detail and led to the discovery of new features, such as the galactic halo and the galactic bulge.

One of the most important discoveries of the 20th century was the realization that the Milky Way is a barred spiral galaxy. This discovery was made possible by new observations and data from the infrared and radio telescopes. The discovery of the bar structure in the center of the galaxy has changed our understanding of the galaxy's dynamics and evolution.

In recent years, the study of the Milky Way has been revolutionized by new data from the European Space Observatory's Gaia satellite. The Gaia mission has provided the most accurate and detailed map of the galaxy to date and has led to new discoveries, such as the existence of a thick disk of stars and the discovery of thousands of new star clusters.

In the 21st century, the study of the Milky Way continues to be an active field of research. Scientists are using new technologies, such as the James Webb Space Telescope, to study the galaxy in even greater detail and to understand its history, structure, and evolution.

The Milky Way galaxy is also a key target for the study of dark matter and dark energy. The galaxy's rotation curve, the distribution of its stars and their motions, and the distribution of its gas, all provide important clues about the nature of these mysterious phenomena.

In recent years, scientists have also used the Milky Way to study the history of the universe. By studying the distribution and ages of stars within the galaxy, scientists have been able to understand how the galaxy formed and evolved over time.

In conclusion, the study of the Milky Way galaxy has a rich history that spans thousands of years. From the earliest observations by ancient civilizations to the most recent data from cutting-edge technologies, our understanding of the galaxy continues to evolve. The Milky Way remains one of the most important objects in the universe to study, as it provides us with a unique opportunity to understand the history, structure, and evolution of our galaxy and the universe as a whole.

Formation of Milky Way Galaxy

The formation of the Milky Way galaxy is a topic of ongoing research in the field of astrophysics. However, current understanding suggests that the galaxy formed through a process known as hierarchical galaxy formation.

This process involves the gradual accumulation of smaller structures, such as gas clouds and dwarf galaxies, through a series of mergers and accretion events.

The initial stages of the Milky Way's formation likely began with the collapse of a large cloud of primordial gas. This collapse was likely triggered by the presence of dark matter, which provided the gravitational force necessary to overcome the pressure of the gas.

As the cloud collapsed, it began to spin and flatten into a disk shape, with the majority of the material concentrated in the central region. Within this disk, small pockets of gas and dust began to collapse to form the first generation of stars.

These first stars were massive and short-lived, and they quickly formed the first generation of supernovae. These supernovae explosions expelled large amounts of gas and dust, which then went on to form the second generation of stars.

As the galaxy continued to grow, it is believed that it merged with many small galaxies, which added to the Milky Way's mass and also contributed to the formation of its spiral structure.

As more material was added to the galaxy, the disk became more unstable, and new stars continued to form in a series of bursts. This process eventually led to the formation of the Milky Way's spiral arms, which are now home to some of the galaxy's youngest and most active star-forming regions.

As the galaxy continued to evolve, the number of stars and the amount of gas and dust in the galaxy increased. This led to the formation of a thick, central bulge and a surrounding halo of old, metal-poor stars.

The exact details of how the Milky Way formed are still not entirely understood, and there is ongoing research to better understand the processes involved. However, it is generally believed that the Milky Way formed through a gradual process of accretion and merging over the course of billions of years.

In summary, the formation of the Milky Way galaxy is thought to have occurred through a process of hierarchical galaxy formation, in which smaller structures such as gas clouds and dwarf galaxies gradually accumulated through a series of mergers and accretion events. This process started with the collapse of a large cloud of primordial gas, and eventually led to the formation of the galaxy's spiral arms, central bulge, and surrounding halo of old, metal-poor stars.

About Solar System

A solar system is a collection of celestial bodies that orbit around a central star, such as our own Solar System which is comprised of the Sun, planets, moons, asteroids, comets, and other objects that orbit around it. Each solar system is unique in terms of the number and type of celestial bodies it contains.

The formation of solar systems begins with a cloud of gas and dust, known as a nebula. Gravity causes the nebula to collapse and form a protoplanetary disk, a flat disk of material that surrounds the central star. Over time, this disk begins to condense and form small particles known as planetesimals. These planet esimals continue to grow through collisions, eventually forming larger bodies known as proto planets.

The proto planets that formed closer to the central star were made mostly of rock and metal, while those that formed farther away were made mostly of ice. The inner rocky planets, also known as the terrestrial planets, include Mercury, Venus, Earth, and Mars. The outer gas giants, also known as the Jovian planets, include Jupiter, Saturn, Uranus, and Neptune. Beyond the orbit of Neptune, there

is a region called the Kuiper Belt, which contains a variety of small icy bodies, including the dwarf planet Pluto.

As the proto planets continue to orbit and collide, they can also form moons. The moons of the solar system can vary greatly in size and composition. For example, Earth's moon is relatively small and rocky, while Jupiter's moon Ganymede is the largest moon in the solar system and is mostly made of ice.

In addition to the planets and moons, solar systems also contain a variety of other objects, such as asteroids and comets. Asteroids are small rocky bodies that orbit in the asteroid belt between Mars and Jupiter, while comets are made mostly of ice and typically orbit in the outer reaches of the solar system in the Kuiper Belt or the Oort Cloud.

In the case of our solar system, it has a central star, the Sun, which is a middle-aged G-type main-sequence star. The solar system also has 8 planets, their moons and other smaller bodies such as comets, asteroids, and Kuiper belt objects.

It's important to note that our solar system is not the only one in the universe. In recent years, scientists have discovered thousands of exoplanets, which are planets that orbit stars outside of our solar system. Some of these exoplanets have been found to orbit within the habitable zone of their star, meaning that they could potentially support liquid water and even life as we know it.

Overall, solar systems are fascinating and complex structures that can contain a wide variety of celestial bodies. They are formed by gravity and other physical processes, and can have a wide range of characteristics depending on the properties of the central star and the materials that formed the planets and other objects. The study of solar systems is an active area of research, as scientists continue to discover new planets and other objects and work to understand the dynamics and potential for life within them.

Types of Solar Systems.

There are several different types of solar systems that have been observed or theorized. Some examples include:

- ➤ Single-star systems: These are the most common type of solar system, and consist of a single star orbited by planets, moons, asteroids, comets, and other objects. Our Solar System is an example of a single-star system.
- ➤ Binary star systems: These systems consist of two stars that orbit each other, and can have planets orbiting one or both of the stars. The orbits of the planets in these systems can be quite complex, as the gravitational pull of both stars affects the planet's orbit.
- ➤ Multi-star systems: These systems consist of three or more stars that orbit each other. Like binary star systems, the orbits of planets in multi-star systems can be complex and affected by the gravitational pull of multiple stars.
- ➤ Circumbinary planets: These are planets that orbit two stars that orbit each other. Such systems were found to exist, for example, Kepler-16b and Kepler-34b
- ➤ Hot Jupiter: These are gas giant planets that orbit very close to their star and have a very short orbital period. They are called hot because they are so close to their star that they can have surface temperatures in excess of 1000 degrees Celsius.

➢ Puffy planets: These are planets that have a much larger radius than expected for their mass. They are thought to have a large gaseous envelope that surrounds a rocky core.

➢ Super-Earths: These are planets that have a mass between 1 and 10 times that of Earth. They are thought to be a common type of planet in the universe.

➢ Rogue planets: These are planets that do not orbit a star, but instead wander through space alone. They are thought to be relatively rare, but may be more common than previously thought.

These are just a few examples of the different types of solar systems that have been observed or theorized. The study of solar systems is an active area of research, and new discoveries are constantly being made.

Our Solar System

Our Solar System is a collection of celestial bodies that orbit around the central star, the Sun. It is comprised of the eight planets, their moons, asteroids, comets, and other objects that orbit around it. The formation of our Solar System began with a cloud of gas and dust known as a nebula. Gravity caused the nebula to collapse and form a protoplanetary disk, a flat disk of material that surrounded the central star. Over time, this disk began to condense and form small particles known as planetesimals. These planetesimals continued to grow through collisions, eventually forming larger bodies known as protoplanets.

The inner rocky planets, also known as the terrestrial planets, include Mercury, Venus, Earth, and Mars. Mercury is the closest planet to the sun, and is small and rocky. Venus is the second planet from the sun, and is similar in size to Earth, but has a thick atmosphere that traps heat, making it the hottest planet in the Solar System. Earth is the third planet from the sun and is the only known planet to support life. Mars is the fourth planet from the sun and is known for its reddish appearance and its potential to support life.

The outer gas giants, also known as the Jovian planets, include Jupiter, Saturn, Uranus, and Neptune. Jupiter is the largest planet in the Solar System and is known for its massive storm, the Great Red Spot. Saturn is known for its rings made of ice and rock particles. Uranus is an ice giant and is known for its tilted axis that causes its seasons to be extreme. Neptune is an ice giant and is known for its strong winds and dark spots on its surface.

Beyond the orbit of Neptune, there is a region called the Kuiper Belt, which contains a variety of small icy bodies, including the dwarf planet Pluto. The Kuiper Belt is also home to many comets, some of which orbit the sun in a region called the Oort Cloud.

The Solar System also has a variety of other objects, such as asteroids and comets. Asteroids are small rocky bodies that orbit in the asteroid belt between Mars and Jupiter, while comets are made mostly of ice and typically orbit in the outer reaches of the solar system in the Kuiper Belt or the Oort Cloud.

The Sun, our central star, is a middle-aged G-type main-sequence star. It is responsible for providing the energy that sustains life on Earth, and its gravity is responsible for holding the Solar System together. The Sun is also responsible for the solar wind, a stream of charged particles that flow out from the Sun and can affect the orbits of the planets and other objects in the Solar System.

The planets in our Solar System also have their own unique features and characteristics. For example, Earth has a diverse range of life forms and a magnetic field that protects the planet from harmful solar radiation. Mars has a reddish appearance due to iron oxide (rust) on its surface, and has the largest volcano and the largest canyon in the Solar System. Jupiter has a massive storm known as the Great Red Spot and has 79 known moons. Saturn is known for its rings made of ice and rock particles and has 82 known moons. Uranus has a tilted axis that causes its seasons to be extreme and has 27 known moons. Neptune has strong winds and dark spots on its surface and has 14 known moons.

Our Solar System also contains a number of dwarf planets, such as Ceres, Pluto, and Eris. Dwarf planets are similar to regular planets in that they orbit the Sun and are large enough to be spherical in shape, but they are not large enough to clear their orbit of other debris.

Elements of our Solar-System.

The elements of our solar system include the Sun, the planets and their moons, comets, asteroids, and other small bodies such as Kuiper Belt and Oort cloud objects, and various types of interstellar dust and gas.

> ➤ The Sun: The center of our solar system and the source of energy for all the planets. It is a medium-sized star and is primarily composed of hydrogen and helium.

- ➢ The Planets: The eight planets that orbit the sun are Mercury, Venus, Earth, Mars, Jupiter, Saturn, Uranus, and Neptune. They have diverse compositions and characteristics, with some being primarily composed of rock and metal (terrestrial planets), and others being primarily composed of gas (gas giants and ice giants)
- ➢ The Moons: Many of the planets have natural satellites or moons. For example, Earth has one moon, while Jupiter has 79 known moons. These moons can be diverse in composition and characteristics.
- ➢ Comets: These are small, icy bodies that orbit the sun. They are made mostly of water ice and dust and when they get close to the sun, they develop a bright coma and a tail of gas and dust.
- ➢ Asteroids: These are small, rocky bodies that orbit the sun. Some asteroids are located in the asteroid belt between Mars and Jupiter while others are scattered throughout the solar system.
- ➢ Kuiper Belt and Oort Cloud objects: These are small, icy bodies that orbit the sun beyond the orbit of Neptune. The Kuiper belt is a disc-shaped region of the solar system that contains many small, icy objects and is thought to be the source of many comets. The Oort Cloud is a spherical region surrounding the solar system that contains many small, icy objects and is thought to be the source of long-period comets.
- ➢ Interstellar dust and gas: These are tiny particles of dust and gas that exist between the stars. They are mostly composed of hydrogen, helium, and other elements such as carbon, nitrogen, and oxygen.

These elements make up the solar system and scientists continue to study and learn about the properties and characteristics of each of these elements to understand the formation, evolution and current state of our solar system.

Formation of Our Solar-System.

The formation of our Solar System is thought to have occurred around 4.6 billion years ago from a cloud of gas and dust known as a nebula. The nebula was composed mostly of hydrogen and helium, with small amounts of other elements such as carbon, oxygen, and nitrogen.

The collapse of the nebula was triggered by a nearby supernova or by the

gravitational pull of a nearby massive object. As the cloud began to collapse, it began to spin faster and flatten into a disk-like shape known as a protoplanetary disk. The central region of the disk began to heat up and eventually formed the Sun.

As the protoplanetary disk cooled, small particles known as planetesimals began to form. These planetesimals were made up of dust and ice, and were thought to have formed through the process of coagulation, where small particles stuck together to form larger ones.

Over time, these planetesimals continued to collide and stick together, forming larger bodies known as protoplanets. These protoplanets were composed mostly of rock and metal, and were thought to have formed in the inner regions of the protoplanetary disk, where it was too hot for volatile materials like ice to survive.

As the protoplanets grew larger, they began to gravitationally interact with one another. This led to a process known as accretion, where protoplanets collided and merged to form even larger bodies. Eventually, these bodies grew large enough to become the terrestrial planets of our Solar System: Mercury, Venus, Earth, and Mars.

In the outer regions of the protoplanetary disk, where it was cold enough for volatile materials like ice to survive, a different process of planet formation occurred. Here, small bodies known as planetesimals began to accumulate into larger bodies known as planetesimals. These bodies were composed mostly of ice and rock, and eventually grew large enough to become the gas giant planets of our Solar System: Jupiter, Saturn, Uranus, and Neptune.

As the protoplanetary disk continued to evolve, leftover planetesimals and debris began to accumulate in the outermost regions of the Solar System, forming the asteroid belt between Mars and Jupiter, and the Kuiper belt beyond Neptune. Some of this debris also accumulated into small bodies known as comets, which are thought to have formed in the outermost regions of the Solar System.

Overall, the formation of our Solar System is thought to have been a complex and dynamic process that involved the collision and accretion of small bodies, the migration of planets, and the influence of the Sun's gravity. The end result is a diverse collection of bodies that make up the Solar System as we know it today.

About Star.

A star is a massive, luminous ball of plasma that is held together by its own gravity. Stars are the building blocks of galaxies and are primarily composed of hydrogen and helium, with small amounts of other elements such as carbon, oxygen, and metals.

The energy that powers a star comes from nuclear fusion reactions that occur in its core. These reactions convert hydrogen into helium, releasing a tremendous

amount of energy in the process. This energy then travels to the star's surface, where it is radiated into space as light and heat.

Stars are classified by their temperature and luminosity, which is determined by the star's mass, composition, and age. The most common type of star is the main-sequence star, which includes stars like the Sun. These stars are relatively stable and spend the majority of their lives burning hydrogen in their cores.

As a star ages, it will eventually run out of hydrogen fuel in its core. When this happens, the star will begin to fuse helium and will expand into a red giant. Eventually, the outer layers of the star will be expelled, forming a planetary nebula and leaving behind a small, dense core known as a white dwarf.

Other types of stars include red dwarfs, which are the smallest and coolest type of star, and blue giants, which are the hottest and most luminous. There are also other types of stars such as neutron stars and black hole which are the end stage of massive star.

Stars play a crucial role in the universe, providing the energy that powers galaxies and the elements that make up planets and life. They also play a key role in the study of the universe, as their properties can be used to determine the age and composition of galaxies and the distance to other celestial objects.

Types of Star.

There are several different types of stars, each with their own unique properties and characteristics. Some of the most common types include:

- ➢ Main-sequence stars: These are the most common type of star and include stars like the Sun. They are relatively stable and spend the majority of their lives burning hydrogen in their cores. They have a wide range of mass, luminosity and size, the lower mass ones are known as red dwarfs and the higher mass ones are called blue dwarfs.

- ➢ Red giants: These are stars that have exhausted the hydrogen fuel in their cores and have begun fusing helium. They are much larger and cooler than main-sequence stars, and are characterized by their deep red color.

- ➢ White dwarfs: These are the remnants of red giant stars that have shed their outer layers. They are extremely dense and are composed mostly of carbon and oxygen.

- ➢ Supergiants: They are the most luminous type of star. They are much larger and more massive than main-sequence stars and can be up to 100 times more luminous than the Sun.

- ➢ Neutron Stars: They are the end stage of massive stars after supernova explosion, they are incredibly dense and composed mostly of neutrons. They have strong magnetic field and emit intense radiation.

- ➢ Black Holes: They are the end stage of massive stars after supernova explosion as well, but they are so dense that nothing can escape from their gravitational pull, not even light. They are invisible, but their presence can be detected by observing their effect on nearby matter.

Each of these types of stars have different life-cycles and eventually end their lives differently, depending on their mass. Understanding these types of stars and their properties helps scientists to better understand the universe and the processes that shape it.

Elements of a Star.

Like the sun, stars are primarily composed of hydrogen and helium, with small amounts of other elements such as carbon, oxygen, and metals. These elements make up the star's atmosphere and interior, and the exact composition can vary depending on the type of star.

➤ Hydrogen is the most abundant element in stars, making up about 74% of the total mass. It is the fuel that powers a star through nuclear fusion in its core, where hydrogen atoms are fused to form helium.

➤ Helium is the second most abundant element in stars, making up about 24% of the total mass. It is the product of nuclear fusion of hydrogen in the core of the star.

➤ Carbon, oxygen, and metals make up the remaining 2% of the total mass of a star. They are formed through nuclear fusion and other processes in the later stages of a star's life.

➤ In addition to these elements, stars also contain various isotopes of elements such as deuterium and lithium.

➤ The temperature and pressure within a star are extremely high, which allows for nuclear fusion reactions to occur. These reactions produce energy in the form of light and heat, which is what makes a star shine.

➤ The energy produced in the core of a star is transported to the surface by convection and radiation, and then radiated into space as electromagnetic radiation.

➤ Some of the larger stars also have different layers such as photosphere, chromosphere and corona with different temperatures and elements.

➤ A Star also have magnetic field which plays a role in the formation of sunspots, flares and coronal mass ejections.

Overall, the properties and elements of a star are determined by its mass, age, and evolutionary stage, and understanding these can help scientists better understand the universe and the processes that shape it.

About Our Sun.

The sun is a star located at the center of the solar system. It is a medium-sized star, classified as a G-type main-sequence star, and is about 4.6 billion years old. The sun is responsible for providing the energy that sustains life on Earth and its planets, and its gravity holds the solar system together.

The sun has several distinct layers, including the core, the radiative zone, and the convective zone. The core is the innermost layer and is the site of nuclear fusion, where hydrogen atoms are fused to form helium. The energy produced in the core is transported to the surface by the radiative zone and the convective zone.

The photosphere is the visible surface of the sun, and it is the layer that gives off light and heat. It has a temperature of about 5,500 degrees Celsius (9,932 degrees Fahrenheit) and is composed mostly of hydrogen and helium, with small amounts of other elements such as carbon, oxygen, and iron.

The chromosphere is the layer just above the photosphere, and it is characterized by its reddish color. It has a temperature of about 10,000 degrees Celsius (18,032 degrees Fahrenheit) and is composed mostly of hydrogen and helium, with small amounts of other elements such as calcium and iron.

The corona is the outermost layer of the sun's atmosphere, and it is characterized by its extremely high temperatures. It has a temperature of up to 1 million degrees Celsius (1.8 million degrees Fahrenheit) and is composed mostly of hydrogen and helium, with small amounts of other elements such as iron and silicon.

The sun also has a magnetic field, which plays a role in the formation of sunspots, solar flares, and coronal mass ejections. These events can have significant effects on the Earth's atmosphere, such as causing auroras and disrupting communication systems.

The sun's energy is generated through nuclear fusion, in which hydrogen nuclei are fused together to form helium. This process releases a tremendous amount of energy in the form of light and heat, and it is this energy that sustains life on Earth.

Overall, the sun is a vital component of the solar system and plays a crucial role in the development and sustainment of life on Earth. Understanding its properties and behavior is essential for understanding the solar system and the universe as a whole.

Formation of Our Sun

The formation of our sun is thought to have occurred around 4.6 billion years ago, through the process of gravitational collapse. This process began with the collapse of a cloud of gas and dust, known as a solar nebula, which was composed mostly of hydrogen and helium, with small amounts of other elements such as carbon, oxygen, and metals.

As the cloud collapsed, it began to spin, creating a flat disk-like structure known as the protoplanetary disk. This disk was composed of the same materials as the original cloud, but the pressure and

temperature in the center of the disk began to increase, causing the hydrogen and helium atoms to fuse together through nuclear fusion. This process released a tremendous amount of energy in the form of light and heat, and it is this energy that powers the sun today.

As the disk continued to collapse, it began to flatten and take on a more circular shape. The material in the disk began to clump together, forming small particles called planetesimals. These planetesimals continued to grow and eventually formed the planets that make up the solar system today.

The process of planet formation was not smooth, there were many collisions, mergers and re-arrangements that happened to form the current solar system.

As the sun continued to contract, it also began to spin faster. This spinning motion caused the sun to flatten out at the poles and bulge at the equator, giving it its characteristic oblate shape.

The sun also has a magnetic field, which is thought to have been generated by convective motions in the sun's interior. This magnetic field plays a role in the formation of sunspots, solar flares, and coronal mass ejections.

Overall, the formation of the sun and the solar system was a complex process that involved the collapse of a cloud of gas and dust, nuclear fusion, and the gradual formation and growth of planets. Understanding the formation of the sun and the solar system is essential for understanding the origins of the universe and the processes that shape it

About Planets.

There are many planets in the universe, but the ones that are most well-known are those in our own solar system: Mercury, Venus, Earth, Mars, Jupiter, Saturn, Uranus, and Neptune. Beyond our solar system, thousands of exoplanets (planets orbiting stars other than our Sun) have been discovered in recent years through various methods such as radial velocity and transit photometry.

These exoplanets can vary greatly in size, composition, and distance from their host star. Some exoplanets have been found to be in the "habitable zone," where conditions could potentially be suitable for liquid water and thus life as we know it to exist. The study of exoplanets is an active field of research with many new discoveries being made every year. Advancements in technology such as the James Webb Space Telescope, set to launch in 2021, will allow scientists to study exoplanets in greater detail and learn more about their potential to support life.

Types of Planets

There are several types of planets that have been classified based on their characteristics and properties. These include:

- ➤ Terrestrial planets: These are planets that are primarily composed of rock and metal and are relatively small in size. The terrestrial planets in our solar system are Mercury, Venus, Earth, and Mars.
- ➤ Gas giants: These are planets that are much larger than terrestrial planets and are primarily composed of gas (primarily hydrogen and helium). They also have a thick layer of metallic hydrogen and a rocky core. The gas giants in our solar system are Jupiter, Saturn, Uranus, and Neptune.

- ➤

➤ Ice giants: These are similar to gas giants but are composed mostly of water, ammonia, and methane. Uranus and Neptune are considered ice giants.

➤ Dwarf planets: These are celestial bodies that are not classified as full-fledged planets, but are larger than most other small Solar System bodies. They are not massive enough to clear their own orbits of other debris. Pluto is the best-known example of a dwarf planet.

➤ Exoplanets: These are planets that orbit stars outside of our solar system. They come in a wide range of sizes, compositions, and orbits and can be classified as gas giants, ice giants, terrestrial planets and even hot Jupiters.

➤ Rogue planets: These are planets that do not orbit any star, instead they float through the galaxy alone. They are thought to be relatively rare.

This classification is not an exhaustive one and scientists are still studying and learning about the properties and characteristics of different types of planets.

Our 8 Planets.

Our solar system consists of eight planets: Mercury, Venus, Earth, Mars, Jupiter, Saturn, Uranus, and Neptune.

➤ Mercury: The closest planet to the sun and the smallest planet in the solar system. It has a heavily cratered surface and is a rocky planet.

➤ Venus: The second closest planet to the sun, it is similar in size and composition to Earth, but has a thick atmosphere that causes a greenhouse effect, making it the hottest planet in the solar system.

➤ Earth: The third planet from the sun, it is the only known planet to support life as we know it. It has a diverse climate and a wide range of ecosystems.

➢ Mars: The fourth planet from the sun, it is often referred to as the "Red Planet" due to its reddish appearance. It has a thin atmosphere, and scientists believe that it may have once had liquid water on its surface.

➢ Jupiter: The fifth planet from the sun, it is the largest planet in the solar system and is primarily composed of gas (primarily hydrogen and helium). It has a thick atmosphere, and its most notable feature is the Great Red Spot, a massive storm.

➢ Saturn: The sixth planet from the sun, it is similar in composition to Jupiter but is slightly smaller. It is known for its rings which are made of ice and rock.

➢ Uranus: The seventh planet from the sun, it is classified as an "ice giant" and is primarily composed of water, ammonia, and methane. Its atmosphere is primarily composed of hydrogen and helium.

➢ Neptune: The eighth and farthest planet from the sun, it is also classified as an "ice giant" and is similar in composition to Uranus. It has a thick atmosphere and a Great Dark Spot, similar to Jupiter's Great Red Spot.

Each of these planets has unique features and characteristics that make them interesting to study. The study of our solar system's planets is an active field of research with many new discoveries being made every year.

Our Moon

The Moon is the Earth's only natural satellite and is about one-quarter the size of Earth. It formed about 4.5 billion years ago from debris left over after a Mars-sized object collided with Earth. The Moon has no atmosphere, and its surface is covered in craters, mountains, and vast plains, with evidence of past volcanic activity. Humans first landed on the Moon in 1969 as part of the Apollo 11 mission. The Moon has been studied and explored for scientific purposes and has played a significant role in many cultures' mythologies and beliefs.

The Moon is composed of several materials including:

- ➤ Silicates (mainly feldspar and pyroxene) making up most of the Moon's crust.
- ➤ Iron-rich oxides found in the Moon's mantle and core.
- ➤ Sodium, potassium, and other light elements, which are scattered throughout the Moon's interior.
- ➤ Oxygen and other volatile elements found in the Moon's soil and rocks.
- ➤ Water ice found in the polar regions, discovered by lunar missions.

It's also worth noting that the Moon has a very weak magnetic field and no atmosphere, making it an inhospitable place for human habitation.

The Moon has a significant impact on the Earth and its inhabitants, both in tangible and intangible ways. One of the most notable impacts of the Moon is its influence on the Earth's tides. The Moon's gravitational pull on the Earth causes tidal movements in the oceans, resulting in predictable patterns of high and low tides. This has significant implications for coastal communities, navigation, and marine life.

In addition to its impact on tides, the Moon plays a role in stabilizing the Earth's axial tilt. This means that the Moon helps to prevent drastic changes in Earth's climate over long periods of time, which is crucial for maintaining a stable environment for life on the planet.

Another impact of the Moon on the Earth is total solar eclipses. When the Moon passes between the Sun and the Earth, it temporarily blocks the Sun's light, creating a total solar eclipse. These events have been observed and studied for thousands of years and have played a role in human culture and mythology.

The Moon has been the subject of mythology and belief systems in many cultures throughout history. In ancient cultures, the Moon was often associated with femininity, fertility, and cyclical patterns of life. The Moon was seen as a symbol of change, with its phases reflecting the changing seasons and life cycles. Many cultures worshipped the Moon as a deity, and it was often the subject of religious rituals and ceremonies.

In modern times, the Moon continues to inspire awe and wonder, and its

presence has been incorporated into art, literature, and music. Despite its impact on the Earth, the Moon remains a mysterious and largely unknown object, with many questions still remaining about its origins, composition, and history.

In conclusion, the Moon has a significant impact on the Earth, affecting tides, stabilizing the axial tilt, and creating total solar eclipses. Its presence has also been the subject of mythology and belief systems in cultures around the world, inspiring awe and wonder for thousands of years.

Apollo11 mission

On July 20, 1969, the United States successfully accomplished one of the greatest feats in human history: landing astronauts on the surface of the Moon. The Apollo 11 mission, commanded by astronaut Neil Armstrong, marked the first time humans had set foot on another celestial body.

The journey to the Moon was the result of a national effort to surpass the Soviet Union in space exploration, and it was driven by President John F.

Kennedy's challenge to the nation in 1961 to land a man on the Moon before the end of the decade. Over the next eight years, NASA worked tirelessly to make this goal a reality.

After a four-day journey from Earth, the Apollo 11 spacecraft entered into orbit around the Moon. Armstrong and fellow astronauts Buzz Aldrin and Michael Collins then embarked on a lunar landing module, named "Eagle," and made their descent to the lunar surface.

As Armstrong took his first step onto the Moon's surface, he famously declared, "That's one small step for man, one giant leap for mankind." The two

astronauts spent several hours on the lunar surface, collecting samples and conducting experiments, before rejoining Collins in orbit and returning to Earth.

The successful completion of the Apollo 11 mission marked a major turning point in human history, demonstrating the capability of humans to explore and conquer the unknown. It sparked a renewed interest in space exploration, and subsequent missions to the Moon continued to advance our understanding of our nearest neighbor in space.

The first human step on the Moon remains one of the defining moments of the 20th century, inspiring generations of scientists, engineers, and space enthusiasts to pursue new frontiers and boldly explore the unknown. Today, the legacy of the Apollo 11 mission continues to drive innovation and exploration in space, as humans set their sights on even more ambitious goals, such as returning to the Moon and exploring the planets beyond our solar system.

Formation & Evolution of Our Moon

The formation and evolution of our Moon is one of the most fascinating and mysterious topics in the study of planetary science. The exact process by which the Moon was formed remains a subject of ongoing research, but there are several widely accepted theories.

One of the most widely accepted theories is the giant impact hypothesis, which states that the Moon formed as a result of a massive impact between the early Earth and a Mars-sized body. According to this theory, the impact caused a portion of the Earth's mantle to be ejected into space, where it eventually coalesced to form the Moon.

Another theory is that the Moon formed from material that was already present in the early solar system and was later captured by the Earth's gravitational pull. This theory, known as the capture hypothesis, has largely been discredited due to evidence that the Moon and Earth have similar isotopic ratios, suggesting a common origin.

Regardless of its exact origin, the Moon has undergone significant changes over its 4.5 billion year history. Early in its history, the Moon was much closer to the Earth and was subjected to intense heat from a period of heavy volcanic activity. This caused the Moon's surface to be covered with a layer of molten rock, which eventually solidified to form the crust.

As the Moon continued to cool, it also shrank in size, causing the formation of large cracks on its surface. These cracks eventually filled with lava, creating the

Moon's distinctive large impact basins, such as the Imbrium and Serenitatis basins.

The Moon's surface has been largely unchanged for billions of years, as the lack of an atmosphere and the Moon's weak gravitational field prevent erosion from occurring. Today, the Moon's surface is covered with a layer of fine dust, known as regolith, which has accumulated over millions of years from the impacts of meteorites and other space debris.

In conclusion, the exact process by which the Moon was formed remains a subject of ongoing research, but the giant impact hypothesis is the most widely accepted theory. Regardless of its origin, the Moon has undergone significant changes over its 4.5 billion year history, including intense volcanic activity, shrinkage, and the accumulation of a layer of regolith. The study of the Moon's formation and evolution continues to provide important insights into the history of our solar system and the formation of other celestial bodies.

Our Earth

The Earth is the third planet from the sun and the only known planet capable of supporting life. It is a terrestrial planet with a diverse environment that includes oceans, continents, polar ice caps, and a thick atmosphere.

The Earth has a diameter of about 12,742 km and a circumference of about 40,075 km. It is the largest of the terrestrial planets and has a mass of about 5.97×10^{24} kg, which gives it a gravitational pull strong enough to keep the Moon in orbit.

The Earth's atmosphere is composed of 78% nitrogen, 21% oxygen, and 1% other gases, including carbon dioxide, which is essential for life on Earth. The Earth's magnetic field protects the planet from harmful solar and cosmic radiation, creating a protective shield known as the magnetosphere.

The Earth's surface is constantly changing due to the processes of plate tectonics, erosion, and volcanic activity. The Earth has a rich geological history, with evidence of life dating back billions of years.

The Earth is home to a diverse range of living organisms, including plants, animals, fungi, and bacteria. The Earth's climate is also diverse, with regions ranging from arctic to tropical.

In conclusion, the Earth is a unique and complex planet, with a rich geological and biological history, a diverse environment, and a protective atmosphere and magnetic field. The study of the Earth continues to provide important insights into the processes that shape our planet and the conditions necessary for life.

The Earth is unique among the planets in our solar system in several ways, including:

➢ Habitability: The Earth is the only known planet capable of supporting life as we know it. This is due to its favorable climate, the presence of water, and a protective atmosphere.

- Liquid Water: The Earth is the only planet in our solar system with liquid water in large quantities. Water is essential for life, and its presence on Earth has played a crucial role in the development of life on our planet.
- Oxygen-rich atmosphere: The Earth has a high concentration of oxygen in its atmosphere, which is crucial for the survival of many life forms. No other planet in our solar system has an atmosphere with a similar concentration of oxygen.
- Strong Magnetic Field: The Earth has a strong magnetic field, which protects the planet from harmful solar and cosmic radiation. This magnetic field is created by the Earth's iron-rich core and is a crucial factor in the habitability of our planet.
- Plate Tectonics: The Earth has a unique geology, with its surface constantly changing due to plate tectonics. This process drives the creation of new land masses, volcanic activity, earthquakes, and other geological phenomena that shape our planet.
- Life: The Earth is home to a diverse range of living organisms, including plants, animals, fungi, and bacteria. This diversity of life is unique among the planets in our solar system and provides important insights into the conditions necessary for life to evolve.

In conclusion, the Earth is a unique and complex planet, with a combination of features that sets it apart from the other planets in our solar system. The study of the Earth continues to provide important insights into the conditions necessary for life and the processes that shape our planet.

Formation &Evolution of the Earth

The Earth is thought to have formed approximately 4.54 billion years ago,

through a process known as accretion. In this process, small particles and objects in the early solar system combined to form larger bodies, eventually leading to the formation of the Earth and other planets.

The early Earth was a hot and violent place, with frequent meteor impacts and intense volcanic activity. Over time, the planet cooled and solidified, with the formation of a solid crust, oceans, and atmosphere.

The Earth's atmosphere initially consisted primarily of methane, ammonia, and water vapor, but as life evolved, it changed to the oxygen-rich atmosphere we have today. The first life forms on Earth are thought to have been simple microorganisms, which evolved into more complex life forms over time.

The Earth's geology has been shaped by a number of processes, including plate tectonics, volcanic activity, erosion, and meteor impacts. Plate tectonics have been particularly important, driving the creation of new land masses and shaping the Earth's surface over time.

The evolution of life on Earth has been a long and complex process, with the first life forms appearing billions of years ago and evolving into the diverse array of life we see today. This process has been driven by a variety of factors, including changes in the Earth's climate, meteor impacts, and the evolution of life itself.

In conclusion, the Earth is a unique and complex planet, with a rich history of formation and evolution. The study of the Earth continues to provide important insights into the processes that have shaped our planet and the conditions necessary for life.

Mass extinction

Mass extinctions are events that have led to the sudden loss of a large number of species over a relatively short period of time. There have been several mass extinctions throughout the history of the Earth, including:

➢ End-Ordovician: Occurring approximately 444 million years ago, this mass extinction led to the loss of approximately 85% of species, including many trilobites and other marine life.

➢ Late Devonian: Approximately 375 million years ago, this mass extinction led to the loss of approximately 75% of species, including many fish and other aquatic life.

➢ End-Permian: The largest mass extinction in the history of the Earth, this event occurred approximately 252 million years ago and led to the loss of approximately 96% of species, including many amphibians and other land-dwelling creatures.

➢ End-Triassic: Approximately 201 million years ago, this mass extinction led to the loss of approximately 80% of species, including many early dinosaurs.

➢ End-Cretaceous: Approximately 65 million years ago, this mass extinction led to the loss of the dinosaurs and many other species, including ammonites, plesiosaurs, and pterosaurs.

These mass extinctions were likely caused by a combination of factors, including meteor impacts, volcanic activity, changes in sea level and climate, and other factors.

In conclusion, mass extinctions have had a significant impact on the evolution of life on Earth, leading to the loss of many species and reshaping the planet's biosphere. The study of mass extinctions continues to provide important insights into the processes that have shaped our planet and the conditions necessary for life.

Formation & Evolution of life on Earth

The formation and evolution of life on Earth is a complex and ongoing process that has taken place over billions of years. The first signs of life on Earth are thought to have appeared approximately 3.5 billion years ago, in the form of simple microorganisms. Over time, these organisms evolved into more complex forms, eventually leading to the evolution of plants, animals, and ultimately, humans.

The early Earth was a harsh environment, with high temperatures and frequent meteor impacts. Despite these conditions, life was able to emerge and evolve, likely due to a combination of favorable conditions, such as the presence of water and organic molecules.

The evolution of life on Earth has been shaped by a number of factors, including changes in the Earth's climate, the availability of resources, and the evolution of life itself. For example, the evolution of photosynthesis allowed plants to produce energy from sunlight, leading to the development of complex ecosystems and the proliferation of life on land.

Over time, the evolution of life on Earth has led to the development of a diverse array of species, each adapted to their specific environment and conditions. Some of the most important events in the evolution of life on Earth include the evolution of multi-cellular organisms, the colonization of land, and the evolution of advanced forms of life, such as mammals and birds.

One of the most significant events in the evolution of life on Earth was the mass extinction of the dinosaurs approximately 65 million years ago. This event led to the proliferation of mammals, which eventually gave rise to humans.

The evolution of life on Earth continues to this day, with new species evolving and existing species adapting to changing conditions. The study of the evolution of life on Earth provides important insights into the processes that have shaped our planet and the conditions necessary for life.

In conclusion, the formation and evolution of life on Earth is a complex and ongoing process that has taken place over billions of years. The study of the evolution of life on Earth continues to provide important insights into the processes that have shaped our planet and the conditions necessary for life.

The theory of evolution by natural selection, proposed by Charles Darwin in his book "On the Origin of Species," is a cornerstone of modern biology. It explains how species change over time through the process of natural selection.

According to Darwin's theory, populations of living organisms tend to produce more offspring than the environment can support. This leads to a competition for resources, such as food and mates. Individuals with traits that give them an advantage in this competition are more likely to survive and reproduce, passing on their advantageous traits to their offspring. Over time, these advantageous traits become more common in the population, leading to the evolution of new species.

Darwin's theory of evolution by natural selection was a major contribution to the understanding of life on Earth. It challenged traditional religious views and provided a scientific explanation for the diversity of life on our planet.

Darwin's theory of evolution by natural selection has been supported by a wealth of evidence, including the fossil record, comparative anatomy, and molecular biology. For example, the fossil record provides evidence of the existence of ancestral species and the gradual evolution of new species over time. Comparative anatomy provides evidence of the common ancestry of different species, while molecular biology provides evidence of the evolutionary relationships between different species.

In conclusion, Darwin's theory of evolution by natural selection is a cornerstone of modern biology and has been supported by a wealth of evidence. It provides a scientific explanation for the diversity of life on Earth and continues to be a major area of study and research in the biological sciences.

Some facts about Astronomy

1. Size of the Universe: The observable universe is estimated to be about 93 billion light-years in diameter. However, the entire universe may be much larger, possibly infinite.

2. Speed of Light: Light travels at a staggering speed of approximately 299,792 kilometers per second (186,282 miles per second). It takes about 8 minutes for light from the Sun to reach Earth.

3. Neutron Stars: Neutron stars are incredibly dense remnants of massive stars that have undergone a supernova explosion. A teaspoon of neutron star material would weigh billions of tons.

4. Pulsars: Pulsars are highly magnetized neutron stars that emit beams of electromagnetic radiation. They rotate rapidly, and their beams are observed as periodic pulses of light.

5. Black Hole Sizes: Black holes can vary in size. Stellar black holes can have a mass several times that of the Sun, while supermassive black holes found in the centers of galaxies can be millions or even billions of times more massive.

6. Auroras: Auroras, such as the Northern Lights (Aurora Borealis) and Southern Lights (Aurora Australis), are mesmerizing light displays caused by the interaction of charged particles from the Sun with Earth's magnetic field.

7. Meteor Showers: Meteor showers occur when Earth passes through streams of debris left by comets or asteroids. As the debris enters our atmosphere and burns up, it creates streaks of light known as meteors or "shooting stars."

8. The Great Red Spot: Jupiter, the largest planet in our solar system, is known for its massive storm known as the Great Red Spot. It has been observed for over 300 years and is larger than Earth.

9. The Oort Cloud: The Oort Cloud is a hypothetical region in the outermost reaches of the solar system, composed of icy objects such as comets. It is believed to be the source of long-period comets that occasionally enter the inner solar system.

10. Dark Sky Reserves: Dark Sky Reserves are protected areas with minimal light pollution, making them ideal for stargazing. These locations offer breathtaking views of the night sky and allow for observations of celestial objects with greater clarity.

Some Important Theories in Astronomy

1. The Big Bang Theory: This theory suggests that the universe began from a hot and dense state approximately 13.8 billion years ago, and has been expanding ever since. It explains the origin and evolution of the universe.

2. General Theory of Relativity: Developed by Albert Einstein, this theory describes gravity as the curvature of spacetime caused by mass and energy. It has been crucial in understanding the behavior of massive objects and the dynamics of the universe.

3. Stellar Evolution: Stellar evolution theory explains how stars are born, evolve, and eventually die. It describes the life cycles of stars, including the processes of nuclear fusion, star formation, and the formation of stellar remnants like white dwarfs, neutron stars, and black holes.

4. Heliocentric Model: Proposed by Nicolaus Copernicus, this theory states that the Sun is at the center of the solar system, with the planets, including Earth, orbiting around it. It replaced the previously held geocentric model, which placed Earth at the center.

5. Nebular Hypothesis: The Nebular Hypothesis proposes that the solar system formed from a rotating disk of gas and dust called a nebula. As the nebula contracted under gravity, it flattened into a disk, and the Sun and planets formed from the material in this disk.

6. Quantum Theory: Quantum theory, also known as quantum mechanics, is a fundamental theory in physics that describes the behavior of matter and energy at the atomic and subatomic scales. It has important implications for understanding the behavior of particles in the cosmos.

7. Cosmic Inflation: Cosmic inflation theory suggests that the universe underwent a period of exponential expansion in the moments following the Big Bang. It explains why the universe appears to be homogeneous, isotropic, and flat on large scales.

8. Dark Matter Theory: Dark matter theory proposes the existence of a type of matter that does not interact with light or electromagnetic radiation but exerts gravitational effects on visible matter. It is thought to make up a significant portion of the universe's mass.

9. Dark Energy Theory: Dark energy theory suggests the existence of a mysterious form of energy that permeates the universe and drives its accelerated expansion. It is responsible for the observed phenomenon of the universe expanding at an increasing rate.

10. Formation of Large-Scale Structures: The theory of the formation of large-scale structures explains how galaxies and galaxy clusters formed from small density fluctuations in the early universe. It describes the growth of structures through gravitational collapse and the formation of cosmic filaments and voids.

About the Author

 Manjeet Barman is a highly accomplished and visionary author, entrepreneur. As the founder and CEO of Scienceoverse, a global scientific platform, he has revolutionized the way science enthusiasts and professionals connect and collaborate. With his strong passion for science and dedication to advancing knowledge, Manjeet has created an invaluable space for individuals from all walks of life to engage in meaningful discussions and contribute to the ever-expanding field of scientific discovery.

Manjeet is an sci-fi writer. With his vivid imagination and curiosity scientific principles, he crafts compelling narratives that blend the boundaries of reality and fiction. Beyond his writing endeavors, Manjeet Barman is also an experienced tutor. He guides and inspires the next generation of scientists and science enthusiasts. Through his tutoring efforts, he imparts a love for science, encourages critical thinking, and nurtures the curiosity that drives scientific exploration.

As an author, entrepreneur, and tutor, Manjeet Barman embodies a remarkable blend of creativity, scientific acumen, and a genuine passion for sharing knowledge. With his new book "The Formation: A Cosmic Journey from nothing to infinity", Manjeet is set to captivate readers and further solidify his reputation as an author and thinker.

Thank you for reading this book...
Send your valuable feedback on: *manjitbarman89@gmail.com*

Make your Notes

Make your Notes

Make your Notes

Make your Notes

Make your Notes

Make your Notes

Make your Notes

Make your Notes

Make your Notes

Make your Notes

Make your Notes

Make your Notes

Make your Notes

End